STRUCTURAL WONDERS OF THE WORLD

PALM ISLANDS

JOY GREGORY

www.av2books.com

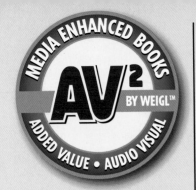

AV² provides enriched content that supplements and complements this book. Weigl's AV² books strive to create inspired learning and engage young minds in a total learning experience.

Your AV² Media Enhanced books come alive with...

Audio
Listen to sections of the book read aloud.

Key Words
Study vocabulary, and complete a matching word activity.

Video
Watch informative video clips.

Quizzes
Test your knowledge.

Go to **www.av2books.com,** and enter this book's unique code.

BOOK CODE

AVS99346

Embedded Weblinks
Gain additional information for research.

Slideshow
View images and captions, and prepare a presentation.

AV² by Weigl brings you media enhanced books that support active learning.

Try This!
Complete activities and hands-on experiments.

... and much, much more!

Published by AV² by Weigl
350 5ᵗʰ Avenue, 59ᵗʰ Floor
New York, NY 10118
Website: www.av2books.com

Library of Congress Control Number: 2019934072

ISBN 978-1-7911-0586-0 (hardcover)
ISBN 978-1-7911-0587-7 (soft cover)
ISBN 978-1-7911-0588-4 (multi-user eBook)
ISBN 978-1-7911-0589-1 (single user eBook)

Printed in Guangzhou, China
1 2 3 4 5 6 7 8 9 0 23 22 21 20 19

022019
103118

Project Coordinator: Heather Kissock
Art Director: Terry Paulhus
Layout: Ana María Vidal

Photo Credits
Every reasonable effort has been made to trace ownership and to obtain permission to reprint copyright material. The publishers would be pleased to have any errors or omissions brought to their attention so that they may be corrected in subsequent printings.

Weigl acknowledges Getty, Alamy, Newscom, iStock, and Shutterstock as its primary image suppliers for this title.

PALM ISLANDS

Contents

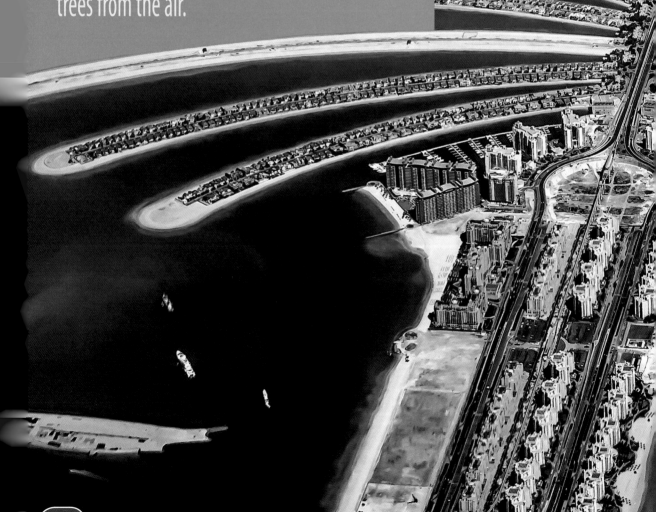

Introducing the **Palm Islands**

The Palm Islands are being developed to look like palm trees from the air.

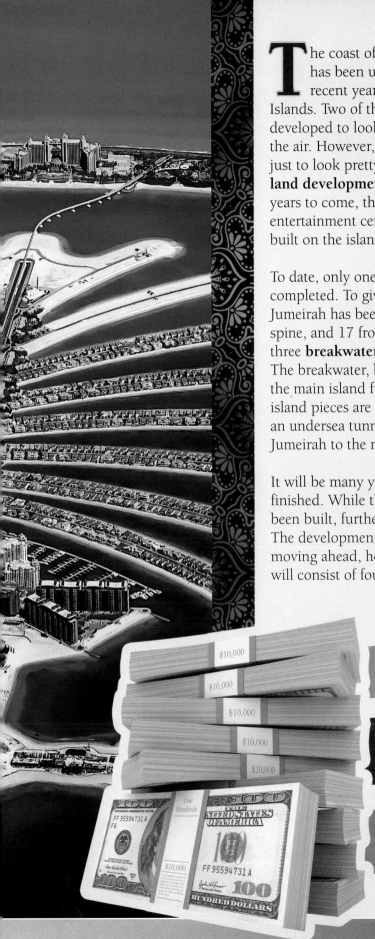

The coast of the United Arab **Emirates** (UAE) has been undergoing a transformation in recent years with the construction of the Palm Islands. Two of these three **artificial** islands are being developed to look like palm trees when viewed from the air. However, the islands are not being created just to look pretty. They are being built to encourage **land development** projects in the UAE. In the years to come, the UAE hopes to see private homes, entertainment centers, and world-class vacation resorts built on the islands.

To date, only one island, Palm Jumeirah, has been completed. To give it the look of a palm tree, Palm Jumeirah has been designed to include a trunk, a spine, and 17 fronds, or leaves. Palm Jumeirah has three **breakwater** islands built around it as well. The breakwater, built in a crescent shape, protects the main island from waves and storms. The various island pieces are linked to each other by a bridge and an undersea tunnel. A separate bridge links Palm Jumeirah to the mainland.

It will be many years before the entire project is finished. While the foundation of Palm Jebel Ali has been built, further construction has been put on hold. The development of Palm Deira, or Deira Islands, is moving ahead, however. When complete, Palm Deira will consist of four small islands connected by bridges.

Palm Jumeirah cost about **$12 billion** to build.

Workers moved **7 million tons (7.1 million metric tons)** of **rock** to build Palm Jumeirah's **breakwater**.

Palm Jumeirah is the **seventh-largest** artificial island **in the world**.

Where Are the Palm Islands?

The Palm Islands are located off the shores of Dubai, the largest city in the UAE. Dubai is also the capital of the emirate of Dubai. The emirate is one of the seven kingdoms that make up the UAE.

MAP OF THE UAE

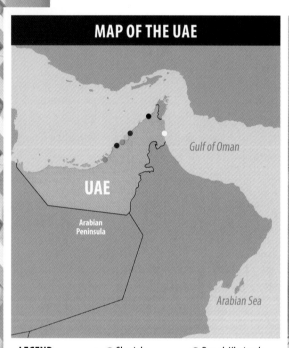

Gulf of Oman

UAE

Arabian Peninsula

Arabian Sea

LEGEND
- ☐ UAE
- ● Dubai
- ● Abu Dhabi
- ● Sharjah
- ● Ajman
- Fujairah
- ● Umm al-Qaiwain
- ● Ras al-Khaimah
- ☐ Land
- ☐ Water

The UAE is a **federation** of seven emirates. These emirates are Dubai, Abu Dhabi, Sharjah, Ajman, Fujairah, Umm al-Qaiwain, and Ras al-Khaimah. Each emirate is governed by its own ruler, with a president responsible for the federation itself. The UAE is located along the east coast of the Arabian **Peninsula**.

Area

Palm Jumeirah created about 1,380 acres (558 hectares) of new land for the city of Dubai. It covers an area approximately 3 miles (4.8 kilometers) in **diameter**.

164 feet (50 m)

Height

The **tripartite** breakwater structure at Palm Jumeirah is 164 feet (50 m) tall when measured from the seafloor.

The UAE is located in an oil-rich part of the world. It started exporting oil to other countries in the 1960s. This led many oil companies to set up offices in Dubai. The emirate became a major economic center for the **Middle East** as a result.

This status, and the wealth it has brought to the emirate, have allowed Dubai to invest in major construction projects. The emirate has developed a reputation for building structures that showcase the UAE's oil wealth and its stature as a major import and export center. The world's tallest building, largest mall, and largest indoor ski slope are all located in Dubai. The city also hosts the world's biggest natural flower garden.

11 miles (18 km) long

650 feet (198 m) wide

Length and Width
The Palm Jumeirah breakwater is 11 miles (18 km) long and 650 feet (198 meters) wide.

Building the Palm Islands

Homes and hotels, including the Waldorf Astoria, were constructed on the breakwater and palm leaves of Palm Jumeirah.

The Palm Islands are being built to **diversify** Dubai's economy. The city is already renowned as a place to see one-of-a-kind structures. The Palm Islands will attract tourists and create a new place for the world's wealthiest people to buy expensive homes on beachfront property.

Dubai's ruler, Sheikh Mohammed bin Rashid Al Maktoum, was the mastermind behind the project. He wanted to ensure that the islands looked natural. Even though they were to be deliberately shaped, it was important to him that the islands blend in with the local environment. For this reason, the builders constructed the islands with as many natural materials as possible. Palm Jumeirah is made mainly of sand. The breakwater has layers of sand and rock.

Ships called dredgers sprayed the sand for Palm Jumeirah into place. The process is called "rainbowing."

Timeline

1990s
Sheikh Mohammed reveals his plan to build three housing and vacation developments off the coast of Dubai.

2001
Construction of Palm Jumeirah's breakwater begins.

2002
Development of Palm Jebel Ali begins near Palm Jumeirah.

2003
Builders announce that the sand foundation of Palm Jumeirah is in place.

2004
Construction of a temporary dam begins at Palm Jumeirah. It holds back seawater so that an undersea tunnel can be built to connect the trunk and the crescent.

2004
Construction begins on Palm Deira. The project is later reduced in size and renamed Deira Islands.

2007
Construction of the Palm Jumeirah tunnel ends, and seawater is pumped back into the area.

2008
Construction of Palm Jebel Ali stops due to a global **recession**.

2019
A commercial area with close to 100 restaurants and 5,000 shops is set to open on Deira Islands.

Palm Islands Features

The Palm Islands development is an ongoing project. Great effort is being made to create a unique environment that will attract both residents and visitors. Much care is being taken to ensure that the islands can survive and thrive in their watery setting.

9 barges, 15 tugboats, 7 dredgers, 30 land-based heavy machines, and 10 floating cranes were used to build Palm Jumeirah's breakwater.

BREAKWATERS The breakwaters at Palm Jumeirah and Palm Jebel Ali contain enough stone to build several Egyptian pyramids. The rock was taken from nearby mines. The Palm Jumeirah breakwater is 34 feet (10.4 m) deep at its lowest point. At low tide, it sits about 13 feet (4.0 m) above sea level.

DEIRA ISLANDS The first two of Deira Islands' four islands will be finished by 2020. When complete, all four islands will add about 6 square miles (15.5 sq. km) of new land to Dubai. The largest of the islands will be attached by bridge to the district of Deira in mainland Dubai. This island will include an $11-million shopping mall and entertainment center.

BEACHFRONT Palm Jumeirah added more than 40 miles (64.4 km) of new beaches to Dubai. This more than doubled the city's original coastline. None of the beaches on the Palm Islands are open to the public. They can be accessed only by the residents of luxury homes and apartments, or by tourists staying at Palm Jumeirah resorts.

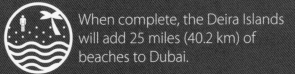

When complete, the Deira Islands will add 25 miles (40.2 km) of beaches to Dubai.

The iconic shapes of Palm Jumeirah and Palm Jebel Ali can both be seen from space.

Palm Islands Structures

The Palm Islands are intended to be a place where people can relax and enjoy beachfront living. The development has been designed to include both traditional and modern elements. Residents and visitors alike can experience the intimacy of island life without giving up the conveniences of mainland living.

ATLANTIS, THE PALM One of the Palm Islands' most prestigious resorts is called Atlantis, The Palm. Located at the top of the Palm Jumeirah breakwater, this five-star hotel reflects the story of Atlantis, a city from Greek **mythology**. The resort has more than 1,500 rooms. This includes the Royal Bridge Suite, which links the resort's two towers. Besides its luxurious rooms, Atlantis, the Palm, also hosts its own tropical gardens, aquarium, and water park.

A stay in the Royal Bridge Suite at Atlantis, the Palm, costs more than $30,000 a night.

BOARDWALK Stretching the length of the island's breakwater, the Palm Jumeirah Boardwalk is a popular place to go for a walk. The boardwalk provides both residents and tourists with beautiful views of Dubai's skyline and coastal waters. There are also opportunities to stop and enjoy the walkway's many food trucks, cafés, and souvenir shops.

The Palm Jumeirah Boardwalk is 20 feet (6.1 m) wide.

The monorail on Palm Jumeirah is 3 miles (4.8 km) long. It cost $380 million to build.

MONORAIL The Palm Jumeirah monorail is the first monorail in the Middle East. A driverless train, the monorail moves people from the Atlantis hotel to Gateway Towers, at the far end of the Palm Jumeirah trunk. Approximately 3,000 people use the monorail every day.

The Grand Design

Construction of the Palm Islands began after many years of planning. Some of the plans have changed as the project evolved. However, the builders remain committed to creating beautiful islands that will not sink or fall into the sea.

Understanding Loads

Builders must take **loads** into consideration when creating structures such as the Palm Islands. There are different kinds of loads. The weight of the structure, including all the materials needed to build it, is called the dead load. Weight from items contained within the structure is called the live load. The kind of soil where the structure is built also affects loads. Soil can sometimes shift or move after construction is completed. The degree to which a structure shifts is called the settlement load. Environmental loads must also be taken into account, especially when building islands. The force of the water pushing against an artificial island impacts how it must be constructed.

Erosion and Compaction

Design of the Palm Islands began at the seafloor. There, sand shifts with the tide. Builders needed to make sure the island's foundation could handle the movement of water and sand, and not shift or **erode**. To decrease the chance of erosion, builders created the foundation using coarse sand from the sea. They believed it could survive the environment better than fine, desert sand. As the sand was being placed, it was "vibro-**compacted**." This process uses vibrations to make the sand settle. Once one batch of sand was compacted, more was added. The process was then repeated. It took eight months to create a stable island at Palm Jumeirah.

The Science Behind the Palm Islands

Scientific expertise and modern equipment were essential to the design and construction of the Palm Islands. Workers depended on both traditional and current technologies to build islands that will stand the test of time.

Pulleys

To build the Palm Islands, construction workers needed to move heavy boulders into place. These boulders served as part of the islands' foundations. The builders used cranes to move the boulders. Cranes use a pulley system to lift objects. Pulleys are a type of simple machine. They are made up of a wheel with a groove along the edge. The groove holds a rope or cable in place. In the case of the Palm Islands project, the cable was attached to a container that held the boulders. The other end was pulled by the crane. The pulley eased the weight of the boulders so that they were easier to lift.

Global Positioning System

Global Positioning System (GPS) technology uses information from satellites that orbit Earth. It was used in the construction of the Palm Islands. GPS pinpointed exactly where dredgers had to spray sand to create the tree-shaped formations. Workers also used GPS to determine how much sand to move to create each section of the design. Crane operators were guided by GPS as well. Divers gave them specific GPS **coordinates** so they could drop breakwater rocks into place.

The Builders

The Palm Islands began with one man's vision for a group of uniquely-shaped artificial islands. To bring his idea to life, companies with highly-specialized workers and equipment were called into action.

Sheikh Mohammed bin Rashid Al Maktoum
Ruler of Dubai

Sheikh Mohammed bin Rashid Al Maktoum was born in 1949. He is the third son of Sheikh Rashid bin Saeed Al Maktoum. Educated in Dubai and England, Sheikh Mohammed was groomed for a role in government from an early age. In 1971, he was named Dubai's minister of defence. Here, he played a key role in directing Dubai's military efforts during various Middle Eastern conflicts. The sheikh was named crown prince of the emirate in 1995. As crown prince, he decided to transform the city of Dubai into one of the world's most luxurious real estate, tourist, and business centers. Construction projects such as the Palm Islands were soon in development. Following the death of his older brother in 2006, Sheikh Mohammed became ruler of Dubai. He also became the prime minister and vice president of the UAE.

Sanjay Manchanda   Chief Executive Officer (CEO), Nakheel

The job of developing Palm Jumeirah was awarded to Nakheel, a land development company in the UAE. The company was founded in 2000. Sanjay Manchanda was appointed its CEO in 2012. He is responsible for the continuing development of the

island. Manchanda has a strong financial background. Prior to his current role, he was a partner with the international accounting firm of Price Waterhouse Coopers. He also worked as Nakheel's chief financial officer.

Pieter van Oord   CEO, Van Oord

Van Oord is a marine construction company based in the Netherlands. Its expertise in offshore **engineering** projects attracted the attention of those building the Palm Islands. This family-run company has been in business for more than 150 years. Before being named Van Oord's CEO in 2008, Pieter van Oord had worked as the company's managing director in the Middle East.

Architects

An architect is a person who designs and plans structures. The master plan for the Palm Islands was created by Helman Hurley Charvat Peacock, an architectural firm based in the United States. Other architectural companies designed specific buildings on the islands. Architects ensure structures meet the needs of the people who use them. They are knowledgeable about the construction techniques and building materials needed to create structures that are safe.

Engineers

Engineers play a key role in construction planning. Before construction begins, an engineer takes the plans for a structure and analyzes how it can be built. More than 1,000 engineers from all over the world have helped build the Palm Islands. Some planned the construction of the islands and breakwaters. Others designed and built the roadways, bridges, and underground tunnel.

Laborers

Many workers are needed to build structures as complicated as the Palm Islands. The project required both skilled and unskilled laborers. Skilled laborers are trained to manage specific tasks. They may operate cranes or install plumbing. Unskilled laborers often do more general tasks. They may move materials, clean job sites, and perform other manual work. At the peak of Palm Jumeirah's construction, 40,000 people worked on the project every day.

Similar Structures

Artificial islands are typically built as a way to add land to a country. Around the world, artificial islands now hold airports, naval bases, homes, and tourist attractions. In Asia, several countries are building islands to ease over-crowding related to population growth. In other areas, people are trying to grow food on floating islands.

Flevoland

Built: 1920–1967
Location: Netherlands, Europe
Design: Cornelis Lely
Description: Flevoland is the largest artificial island in the world. The twelfth province of the Netherlands, it covers 931 square miles (2,411 sq. km) of land. This land was reclaimed by building **dikes** to keep water off the land. About 400,000 people live in Flevoland. Agriculture and tourism are its main industries.

Yas Island

Built: 2006–2009
Location: Abu Dhabi, UAE
Design: Benoy (Architects)
Description: Yas Island is the second-largest artificial island in the world. Once part of mainland Abu Dhabi, it was created by digging a **canal** between the mainland and the 6,200 acres (2,509 ha) of land that form the island. The island was designed to be a hub of entertainment and leisure activity. It includes a Grand Prix racecar course and Abu Dhabi's largest shopping mall.

Kansai International Airport Island

Built: 1987–1994
Location: Osaka, Japan
Design: Renzo Piano
Description: Kansai International Airport Island is the world's third-largest artificial island. Created to relieve the congestion at Osaka's airport, Kansai cost more than $18 billion to build. The island is located 3 miles (4.8 km) offshore. It is accessible by bridge and train. The airport handles 25 million passengers a year.

Lantau Tomorrow Vision

Built: 2025–2032
Location: Hong Kong, China
Design: To be announced
Description: A shortage of space makes Hong Kong one of the world's most expensive housing markets. Lantau Tomorrow Vision is meant to ease the problem. This group of artificial islands is expected to create 4,200 acres (1,700 ha) of new land for the city. The land will be used to build up to 400,000 new homes. More than 1 million people might eventually live at Lantau. The development will take two or three decades to complete.

At Issue

A project the size of the Palm Islands does not come without problems. The impact the project would have on the environment, both during and after construction, was always a concern. However, the environment has also affected the Palm Islands, endangering their shape and structural integrity.

WHAT IS THE ISSUE?

Sediment stirred up by the construction of Palm Jebel Ali decreased the amount of light and oxygen in the water.

A strong, hot wind called a "shamal" blows over the region several months a year.

In the first six months of 2018, 8.1 million people came to visit the Dubai area. Dubai already has a population of 3.1 million.

EFFECTS

The corals that lived in these waters were not receiving the nutrients they needed to survive.

The shamal heats the water and strengthens currents. This can cause the palm-tree shape of the shoreline to change.

The additional people are causing an increase in sewage, garbage, and other forms of pollution.

ACTION TAKEN

The corals were relocated to cleaner waters northwest of the island.

The shorelines of artificial and natural beaches are monitored. New sand is added as needed to maintain the shape.

Research studies are being undertaken to determine how to reduce pollution that is caused by an increase in population.

Activity

Erosion on the Beach

Water currents can have a significant impact on a shoreline, causing it to erode and change shape. To see how this happens, make a model of your own beach. Study how the shoreline is affected when you change the current.

Materials

Timer

Paper

Pencil or pen

5 cups (1.2 liters) of sand

2 cups (0.5 L) of aquarium gravel

6 cups (1.4 L) of water

Empty water bottle, sealed with lid

Paint roller tray

Instructions

1. Pour the sand into the paint roller tray.

2. Build a beach by pushing the sand into an even pile in the shallow end of the tray.

3. Place the aquarium gravel to one side of your beach.

4. Slowly pour the water into the well at the deep end of the try. This is your ocean.

5. Notice where the beach and water meet in the tray. Sketch a picture.

6. Float the water bottle in the tray. Place it horizontally so that it is parallel with your beach.

7. Set the timer for 60 seconds. Use your fingers to create currents in the water and make the bottle move up and down.

8. When the timer is done, remove the water bottle and look at your beach.

9. Record your observations:

    a. How has your shoreline changed?

    b. How did the currents change the beach where the aquarium gravel was placed?

    c. Draw a picture that shows the changes.

8 Ways to Test Your Knowledge

1
In which country is the city of Dubai located?

2
Which leader of Dubai was the mastermind behind the Palm Islands project?

3
How many tons (metric tons) of rock were used to build the breakwater at Palm Jumeirah?

4
How many breakwater islands does Palm Jumeirah have?

5
What parts of Palm Jumeirah are linked by an undersea tunnel?

6
How wide is Palm Jumeirah Boardwalk?

7
Which company developed Palm Jumeirah?

8
How much did it cost to build Palm Jumeirah?

Answers

1. UAE **2.** Sheikh Mohammed bin Rashid Al Maktoum **3.** 7 million tons (7.1 million metric tons) **4.** Three **5.** The crescent and trunk **6.** 20 feet (6.1 m) **7.** Nakheel **8.** About $12 billion

Key Words

artificial: constructed by humans, not by a natural process

breakwater: a barrier that breaks the force of waves

canal: an artificial waterway that has been made for boats to travel along

compacted: firmly packed

coordinates: a set of numbers that indicate an exact position

diameter: the longest distance across a circle

dikes: walls used to hold back water

diversify: to increase variety

emirates: countries or states ruled by monarchs called emirs

engineering: using scientific principles to design and build machines and structures

erode: to gradually wear away

federation: an organization made up of smaller groups, parties, or states

land development: the conversion of raw land into construction-ready building sites

loads: the collection of forces acting on an object

Middle East: the countries of southwestern Asia and northern Africa

mythology: traditional stories about gods and heroes

peninsula: a piece of land that is nearly surrounded by water but still connected to land

recession: a period of decreased economic activity

sediment: material, such as sand and soil, that settles on the bottom of the seafloor

tripartite: made up of three parts

Index

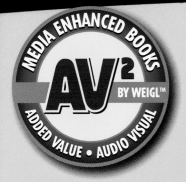

Log on to www.av2books.com

AV[2] by Weigl brings you media enhanced books that support active learning. Go to www.av2books.com, and enter the special code found on page 2 of this book. You will gain access to enriched and enhanced content that supplements and complements this book. Content includes video, audio, weblinks, quizzes, a slideshow, and activities.

AV[2] Online Navigation

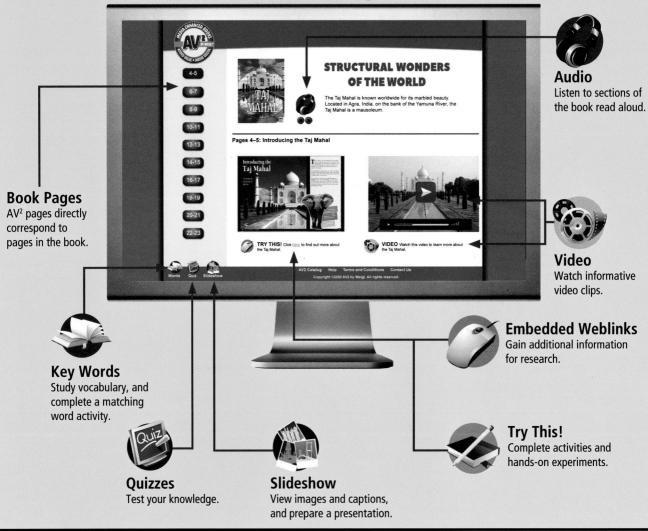

Book Pages
AV[2] pages directly correspond to pages in the book.

Audio
Listen to sections of the book read aloud.

Video
Watch informative video clips.

Key Words
Study vocabulary, and complete a matching word activity.

Embedded Weblinks
Gain additional information for research.

Quizzes
Test your knowledge.

Slideshow
View images and captions, and prepare a presentation.

Try This!
Complete activities and hands-on experiments.

AV[2] was built to bridge the gap between print and digital. We encourage you to tell us what you like and what you want to see in the future.

Sign up to be an AV[2] Ambassador at www.av2books.com/ambassador.

Due to the dynamic nature of the internet, some of the URLs and activities provided as part of AV[2] by Weigl may have changed or ceased to exist. AV[2] by Weigl accepts no responsibility for any such changes. All media enhanced books are regularly monitored to update addresses and sites in a timely manner. Contact AV[2] by Weigl at 1-866-649-3445 or av2books@weigl.com with any questions, comments, or feedback.